EARLY BRIDGES OF SUFFOLK

Carol Twinch

By the same author

Women on the Land: Their Story During Two World Wars
In Search of St Walstan: East Anglia's Enduring Legend
Tithe War 1918-1939: The Countryside in Revolt
Great Suffolk Stories
Saint with the Silver Shoes: The Continuing Search for St Walstan
Bury St Edmunds: A History and Celebration
Walk Around Historic Bury St Edmunds
Ipswich: Street by Street
Little Book of Suffolk
The History of Ipswich
Essential Essex
Walks Through History: Ipswich
The Norwich Book of Days
Saint Walstan: The Third Search

Plain and Simple Egg Production
So You Want to Keep Sheep
Poultry: A Guide to Management

First Edition 2016

Published and Printed by
Leiston Press Ltd
Masterlord Industrial Estate
Leiston
Suffolk
IP16 4JD
Telephone Number: 01728 833003
Web: www.leistonpress.com

ISBN 978-1-911311-15-7

© Carol Twinch

All rights reserved. No part of this book may be reproduced, stored in a retrieval system, or transmitted in any form or by any means electronic, mechanical, photocopying, recording or otherwise, without the prior permission of the author.

CONTENTS

Introduction   1

Chapter 1   3
*Roman fords – Anglo Saxon meeting and crossing places – early
wooden bridges – early medieval will bequests – the Norman
Conquest - drawbridges*

Chapter 2   13
*Medieval bridges – Magna Carta – the common burdens - will
bequests – stone replaces wood – the Established Church as
bridge builders – early tolls - ferries - packhorse bridges*

Chapter 3   32
*Bridge hermits – medieval bridge chapels - tolls – stone crosses
at bridges - the Dissolution of the Monasteries*

Chapter 4   49
*Plantagenet will bequests - Bridges Act 1530 – town
benefactors – moat bridges – Tudor bridges - Christopher Saxton
and John Speed maps – Robert Reyce – introduction of
turnpike trusts*

Bibliography   60

The gauging station at Knettishall Heath bridge is looked after by the Environment Agency as even a small amount of vegetation could affect the flow data. Pictured are the Agency's Andy Allum and Alan Bidwell.

Front cover: Cattawade

INTRODUCTION

Most Suffolk villages and towns have roots stretching back to pre-history and a majority can lay claim to at least Roman or Anglo Saxon habitation. It made sense for the earliest people to live beside a ready supply of running water and, since the county is marbled with rivers and streams, there are a large number of riparian hamlets, parishes and towns in Suffolk. There were spiritual dimensions to living close to water and it is the considered opinion of topographers and historians that all early settlements were normally within a mile of a watercourse.

Once people became rooted in a particular spot they needed to trade and travel beyond the parish boundaries, at first locally and then further afield. They employed an incalculable number of fords and crossings to go hither and yon and until 19th century technology interfered, these routes became entrenched as highways that were followed over the centuries. The best-used ones progressed from fords to bridges of increasing permanence and durability and communities grew up around them.

Those historic bridges that are left to us are works of art and reflect the age in which they were built and those who designed them. Gaze in wonder on the remnants of the Abbot's Bridge in Bury St Edmunds, or at the warm red bricks of the double bridges at Chelsworth and Ufford, the graceful lines of the bridge at Brent Eleigh or the simple beauty of hundreds of small but functional bridges across the county: together they embody the evolution of travel and human communication.

There are few road bridges upon which you can safely stand and watch the water rush, trickle or meander under your feet, but Bourne Bridge in Ipswich is one and Homersfield Bridge is another. Hempyard Bridge at Ixworth is a delight and the Old Brook Bridge just outside Walsham le Willows is pedestrianised and adjacent to its busy and noisy replacement. Melford Bridge at Thetford is another where the traffic roars past but the bridge itself is only passable by pedestrians and cyclists.

Suffolk people can generally point to somewhere they live or have lived that has a river bridge nearby: my ancestors lived in Rickinghall during the 15th century and I was born in Eye, which has several bridges one of which I must have crossed soon after birth. My formative years were spent in a once-moated farmhouse at Hoxne, a village with several bridges including the famous Goldbrook Bridge, and I remember my mother driving us across what was the county's last surviving toll bridge at neighbouring Syleham in the 1950s.

Summer holidays were spent with my grandparents at Brook Farm, Charsfield, which takes its name from Potsford Brook that has a small bridge near the house and the brook runs alongside the farm meadows. For over 15 years I lived at Bawburgh in Norfolk which sits astride the River Yare that has a long bridge history and its own bridge hermit. Back to Suffolk again, Sternfield this time, where for two years I passed daily over the Benhall Green ford; and now Rendham where the bridge goes over the River Alde.

This book sets out to highlight the historical river bridges in the county from the early ford crossings to Tudor and Elizabethan times. It is not a history of bridge making; nor is there room for more than a few of the uncountable moat bridges, ford crossings and watersplashes (or splashes as they are known locally). Footbridges, which must number in the thousands, are mentioned in passing; similarly carnsers, causeways and Irish bridges. No one is quite sure why Irish bridges are so named but they are raised walkways alongside fords.

The story of Suffolk's bridges did not end, of course, in the 16th and 17th century and Volume Two will bring the story up to date. The Orwell Bridge, for example, carries upwards of 50,000 vehicles a day over the River Orwell and is perhaps the best known of Suffolk's multitude of modern bridges. The streamlined Lemon Hill Bridge at Tattingstone and Sudbury's Ballingdon Bridge are also significant milestones in the evolution of Suffolk bridges that began in pre-history and evolved into the glorious mix in evidence today.

Many thanks to all the kind and helpful people we met on our year-long foray in search of Suffolk's bridges, and thank you as always to the other half of 'we', my husband Christopher.

Carol Twinch
Rendham 2016

CHAPTER 1

East Anglia was itself a kind of bridge, way back in prehistoric times, being the final piece of Europe to break away from the rest of the continent, around 6,500 BC when increasing temperatures melted the ice caps. Gradually the land bridge to Europe disappeared under rising sea levels and Britain became an island. Now, of course, there are no bridges of any kind to the east out of Suffolk or Norfolk and the only connection between Britain and mainland Europe is the Channel Tunnel. Instead, bridges lead out of the county into Norfolk to the north, Essex to the south and Cambridgeshire to the west.

Suffolk does not have grand multi-arched bridges spanning huge rivers because there are no vast rivers in the county. The land is generally flat and there is no run-off from mountains to congregate in valleys and carve out rivers like the Tyne, the Dee or the Severn. Nor is there an extensive catchment area, like the Thames; the east of England is the driest part of the British Isles and the rivers are many but modest. True the Orwell Estuary is wide but the river channel itself winds inland making it difficult for shipping to get easily up to Ipswich with high tide hiding the extensive mud banks.

The county's smaller rivers gave rise to smaller communities but because of the relatively gentle topography East Anglians have always moved easily across the land. There was often a shallow place in a river bed that offered the opportunity for a ford.

The earliest bridges were wooden, probably no more than a few logs placed side by side across the lowest and hardest part of a watercourse, but for generations river crossings were predominantly fords.

There was one point on the River Deben still locally called **Hawkeswade** (near **Ufford** Bridge) where you could wade across the river at low tide. The same point later became Uffa's ford, which gave Ufford its name and confirms this as an ancient river crossing. Uffa, the 6th century King of East Anglia and a member of the Scandinavian Wuffings dynasty, had his 'seat' at Rendlesham and to gain access to the western bank and beyond he would have needed to cross the river at its nearest crossable point, in all probability following a much earlier track way or footpath.

This is one of the many fordings which evolved into traditional routes, as evidenced today by the twin semi-circular bridges at **Ufford** taking the place of what were in ancient times three separate fords, alongside each other.

Cattawade (first crossing on the River Stour) is thought to have another meaning – 'a ford frequented by wild cats' - but could as easily indicate an early wading point for humans always assuming the wild cats were elsewhere.

In the 1st century there was a fording place on what is now the River Gipping known as **Horsewade** which is taken to mean a place that a horse could wade across. It later gave its name to a mill and appears in documents as early as AD970. The nearby Handford crossing in **Ipswich** indicates, by its very name, that here was a solid ford (Old English *han* meaning 'stone' or 'rock') and is likely to be near or at Horsewade.

A hamlet of **North Cove** named **Wade** is usually interpreted as being a ford that could be waded across the words 'ford' and 'wade' being almost interchangeable. Wade Hall is the only reminder of it.

Even a brief glance at an Ordnance Survey map shows fords dotted all over the place and there are over twenty Suffolk parish names that end in 'ford'. Place names are a give-away and almost any with 'ford' in it is self-evident: **Brockford** (ford over the brook), **Billingford** (the ford of the *Billingas*), **Chillesford** (gravel ford), Culford (*Cula's* ford), **Fordley** (clearing or wood by a ford), **Kentford** (ford over the River Kennett), **Lackford** (ford where leeks grew) and **Yoxford** (a ford that could be passed by a yoke of oxen) to name but a few.

Thelnethham has a ford and a bridge over the Little Ouse river and it must have had one of the latter for a very long time since the name of the village derives from the Old English 'thel' which means a plank bridge; 'elfitu' means swans and 'hamm' indicates a meadow or enclosure. Hence Thelnethham is the meadow with the plank bridge frequented by swans.

The bridge at **West Stow** has a long pedigree of occupation as there is archaeological evidence that the fording place was settled in the early and late Neolithic, Iron Age and early Anglo Saxon periods.

People gathered at **Playford** and **Glemsford** for games, sports and merriment and both fords survived to become bridges. A 15th century packhorse bridge is said to have once spanned the river at Playford but there is no trace of it now.

Some crossings were named for local Anglo Saxon individuals rather than a geographical feature but if the site did not evolve into a regular route, and a bridge built, its name was lost along the way. Some, though, had to wait a long time for a bridge in spite of popular usage. **Combs Ford,** a drovers' ford on the route from Bury St Edmunds over the Rattlesden River to Stowmarket, had to wait until 1755 for its first bridge.

The words 'bridge', 'brook' and 'ford' are applied to farms, cottages, roads and streets all over Suffolk but **Bridge Street,** a hamlet of **Alpheton** on the A134, has the distinction of being one of only a few Suffolk place-names to retain the word 'bridge'. The hamlet is also peculiar in that the northern part lies within the parish of Alpheton but the southern part lies within the parish of Long Melford. The bridge crosses Chad Brook and is close to Ford Hall, which would indicate a traditional fording place.

Settlements grew up beside fords and gave their names to crossings sometimes known as the 'watersplash' or even the 'tidal road', imbuing them with an almost spiritual quality. The most famous watersplash, or splash as it is called in Suffolk, is that at **Kersey** where a tributary to the River Brett runs across the village highway. Another, at **Grundisburgh,** is hard-bottomed and crosses the village green where it is the frequent haunt of ducks. The ford at **Cookley** has a well-used raised footbridge (sometimes called an Irish bridge) as in winter the water can rush across the road in a mini torrent.

Due to the lack of natural stone in the county there is no record of any discernible ancient stepping stones in Suffolk but no doubt there were make-do steps – or even the odd clapper-type bridge – where enough stone could be found. These clapper bridges were the first simple stone bridges which consisted of stone slabs resting on stone piers over a small stream.

The best example of historic stepping stones is in **Ipswich** where Stepples Street (known today as Orwell Place) took its name from the raised stepping stones. An overflow from a spring ran down Bolton Lane and into the stream which flowed along Upper and Lower Orwell Street and was often difficult to cross, thus the area became known as the 'stepples'. The stones were gone by 1796 when the area was paved as part of a town improvement programme.

At **Debenham**, which incidentally has the longest ford in England, stepping stones of sorts were constructed of bricks laid across the infant Deben and used regularly before the first bridge was built, as recently as 1800. The bricks were often submerged and were removed when Market Bridge was constructed to take carriages along the Turnpike from Woodbridge to Eye. Unfortunately that bridge was so badly built that in 1808 it had to be taken down and reconstructed at a cost of twenty pounds.

An exception to the no-stone rule are the sarsen stones of **Boxford** one of which is seen at the base of a wall next to the bridge. Another has been placed at one end of the bridge and there are several small ones in the Box river bed. As these sarsens are generally thought to be naturally occurring they may once have been used to strengthen the ford bottom. Some of the larger ones are thought to have been marker stones for the edge of the old ford.

In spite of there being a dearth of suitable stones there were bound to have been fords that were built up with available flint or rubble as evidenced by several mentions in medieval wills to repairs to the local 'causey'. The term causey usually indicated a pathway across wet or soft terrain, and is better associated with the packhorse era, but such crossings are sure indicators of ancient travel routes. These causeys were particularly prone to being washed away during the winter and required constant maintenance, but many existed well on into the 15th and 16th century.

In 1487 William Church of the Priory of **St Olaves** left money for the causey to be repaired with posts (*stulpio*) and planks (St Olaves was in Suffolk until 1972). The widow Mariota Sprunte of Clare also left 6 shillings and 8 pence in her will of October 1488 'to the reparation and emendation of Chilton causey and the foot bridges of the same'. Richard Haddenham of **Kentford** - who died in 1542 - also left 6 shillings and 8 pence 'to the reparation of the causey from Kentford Bridge to my door'. The latter shows that causeys remained in use long after a bridge was built.

The Romans built wooden bridges, and probably stone ones, but none remain in Britain let alone Suffolk although there are a few apocryphal Roman associations. There was once a bridge on the Stoke Road leading out of **Clare** which used to be called Rome Bridge and the so-called packhorse bridge at **Kentford** (contemporary with the 15th century Moulton Bridge) that was washed away in the 1970s was known as the Old Roman Bridge, where there was 'always' a ford nearby.

Stoke Ash is believed to have been a Roman military trading post and the remains of a timber bridge there have been suggested as a contemporary crossing of a tributary of the River Dove.

All Stratfords, the ford of the *Via Strata,* are on Roman roads and **Stratford St Andrew** marks the fording of the River Alde by the Romans, i.e. 'the ford by which a Roman road crossed a river'. Stratford St Andrew is one of only thirteen Suffolk place names that appear on the Gough Map drawn up around 1360. It is kept at the Bodleian Library in Oxford. The Gough is thought to be the earliest attempt at a road map of Great Britain and is the centre of much academic attention, some scholars believing it to date even further back, possible to the reign of Edward I (1272-1307). **Cattawade** (then 'Catiwad') is one of the thirteen, as is **Brandon** ('brandofery').

Also Roman in origin is **Stratford St Mary,** across the River Stour on the Essex border, which had an early crossing and may have replaced an even earlier one closer to Dedham at the northernmost bend of the river. When the Romans left

Britain in around 450AD many of their roads, including the fords, ceased to be used but the Stour crossings retained their importance in post-Roman Suffolk.

The bridge at **Glemsford** – or rather the original ford over the River Glem – was said by the Glemsford historian, Revd Kenneth Glass, to be on a path leading to the Via Devana, a military Roman road that ran from Colchester to Chester. It is believed there was a small Roman establishment at the ford where the troops exchanged horses. Kenneth Glass further suggested that it served as a 'police detachment' to protect the ford against the raids of the native Iceni tribe.

The Romans had a settlement at **Withersfield** where the road from Cambridgeshire leads into the western edge of Suffolk. In the 18th century a Roman cemetery was discovered at Withersfield's Melbourn Bridge. It is likely that they made use of the ancient river crossing to travel between their many villas in the area.

Where the hump-backed **Coddenham** Bridge now straddles two tributaries of the River Gipping the Romans had a station called *Combretonium*. As it was an important administration and transport centre it is likely that they did not make do with a ford but threw an early bridge of some kind over the streams.

Then there is the myth of a Roman bridge, or hard, that once crossed the Orwell south of **Ipswich.** The Bridge Wood in Orwell Country Park reputedly takes its name from this wooden bridge at the 'old Roman crossing point'. It is true that at low tide a straight stretch of shale and stone can be seen in the mud, which is considered to be the remains of a low tide crossing and sometimes referred to as Downham Bridge. But during the extensive archaeological survey carried out prior to the construction of the Orwell Bridge in the 1970s and 80s, no evidence was found for what has become known as the 'urban myth of the Roman bridge'.

For evidence of early fords there are occasional archaeological pointers although there are few bridge stories concerning the 10th century Danish incursions into Suffolk. However, during 1651 at the bridge in the ancient hamlet of **Brockford** (now Wetheringsett-cum-Brockford) the skeleton of a 'mighty giant' was discovered. He was ten feet tall, his leg bone was 'about the width of a middling woman's waist', and there were eleven 'huge' teeth among the bones. The roadmen were digging the 'gravelly way' when the skeleton was unearthed and many thought him to be a Dane. He was said to have been buried in line with the road, his head pointing towards Ipswich, which in 991 was ravished by Danish raiders.

Workmen digging near one of the bridges at **Debenham** discovered a less spectacular find, but one nonetheless that would indicate the site to be an ancient crossing. The object was a coin of the realm of Harold I, king of England from 1035-1040. Harold claimed to be the youngest son of the Danish King Cnut and was known by his nickname as 'Harefoot' to celebrate his ability to run fast and prowess in the hunting field.

If the legend concerning St Edmund hiding under Goldbrook Bridge in **Hoxne** is to be believed, a bridge was there in AD869 and thus one of the earliest bridges in Suffolk. The story goes that King Edmund was in retreat from the Danes and took temporary refuge under the bridge. By chance a wedding party was crossing the brook and half way across the bride looked down at the water and started as the metal from Edmund's spurs glinted in the sunlight and reflected on the surface. The couple gave the game away and before long a party of Danish soldiers hastened to Hoxne where they arrested Edmund, tied him to a tree and fired arrows at him in an attempt to persuade him to renounce Christianity. Having failed they decapitated him and flung his head into a nearby wood, where it was guarded by a wolf until rescued by local monks. The body was taken to a nearby chapel and Edmund was proclaimed a saint and martyr. Eventually the body was taken to Beodricsworth (Bury St Edmunds) but not before St Edmund was said to have put a curse on those who cross the bridge on their way to get married.

In 1712 Edward Steele, author of Parish Notes, visited Hoxne. After enquiring about the legend of Goldbrook Bridge he was shown a small bridge under which Edmund hid from the Danes. Steele was told that crossing it was perilous, especially for those 'newly entered into matrimonial honours', who would go some miles out of their way to avoid using the bridge.

In fact this may not be the cursed bridge at all, since there were always four or more bridges in Hoxne, one of them named Rase or Rays Bridge. If or when Rase was renamed Goldbrook is open to discussion.

In the vicinity of the Nuns Bridges at **Thetford** (in Suffolk until 1889 though now mostly in Norfolk) is a rival to Hoxne as the site of St Edmund's capture and martyrdom by the Danes. In AD869 the Danish invaders laid siege to Thetford which stood on the Icknield Way at the crossing point of the Rivers Little Ouse and Thet. A battle took place and Edmund was reputedly captured and killed thereabouts.

Another of the tales and legends about St Edmund is one which concerns the **Barnby** crossing. While being pursued by the Danes, Edmund escaped across the Waveney using this ford. Called 'Berneford' it was known only to him and he used it successfully to elude the marauding Danes. Some time later he re-grouped his army and on a spot between the ford and Carlton Colville engaged the Danes in battle and was victorious.

Whether or not the two legends are linked is unknown, but Barnby Bridge is reputedly haunted by one of Suffolk's several Black Shucks. Shucks were reputedly large black dogs with flaming eyes and shaggy black hair that belong to the Devil and roam the countryside spreading terror wherever they went. Their howling was said to make the blood run cold. If the Barnby Bridge Shuck followed you, a close relative would die soon after.

Another Shuck is said to haunt two bridges along Mill Road in the Sandy Hills area of **Wissett** and are known as the Bridge Dogs. It is likely that many stories of a ferocious Black Shuck were put about by smugglers in the 18th century, to keep people away from certain areas where many of their nefarious activities took place.

In Anglo Saxon times river crossings were invariably where the moot was held, when the leading or prominent men of a locality who met to discuss important matters.

Mutford, for example, takes its name from the Old English *Motford* - 'ford at which moots were held'. The moot, or legislative assembly, took place on the bank of the Hundred River that flows between the Hundreds of Mutford and Blything. The river gives its name to the Hundred River Bridge only a mile or two away at **Ellough.**

The Mutford crossing at Oulton was an ancient crossing-place between Lake Lothing and **Oulton Broad** which at one time was nothing more than an earth dam or causeway.

Anglo Saxon moots were also held at **Wilford Bridge** which is the lowest crossing point on that stretch of the River Deben. The moot was not only to discuss local affairs but its members dispensed justice: hangings took place a little way up from the river on part of what now forms the golf course. It is only a very short distance away from the Anglo Saxon burial grounds at Sutton Hoo discovered in 1939. The ship burial of the 7th century warrior king, together with his most treasured possessions, caused a seismic re-think on 6th and 7th century history. The revelations of the Sutton Hoo burials were and remain unparalleled in Suffolk archaeology.

Wilford was an early crossing of the Deben and there have been many bridges over the years. The current bridge was built in 1936 to replace an earlier one probably dating to around 1554 or 1575 that no doubt replaced an even older one, and so on. Piers from an old bridge can be seen at certain low tides.

Down river from Wilford was another Anglo Saxon crossing place over the King's Fleet near the mouth of the River Deben. It was called **Gosford** (Goseford), a ford haunted by geese. In the 13th century Gosford port was known for its imports of German and French wine and for its victualling services to ships mustering or sheltering in the Gosford haven. Although it carries an Anglo Saxon name the area was known to the Romans who may well have laid the foundations for the ford. Gosford is one of the few places marked on the earliest map of Suffolk that was drawn between 1173 and 1200.

One of the oldest commercial river crossings in Suffolk, if not the oldest, is the Stoke Crossing in **Ipswich**. It is likely that there were two bridges here, built at different times, one of which later retained the name North Bridge for a while. One or both bridges had a barrier at one end under the control of a bailiff.

Suffolk, and particularly the ports of Dunwich and Ipswich, prospered since before Anglo Saxon times principally because of their proximity to European trade routes. From around the second decade of the 8th century the manufacture on an industrial scale of Ipswich Ware pottery necessitated routes out of the town to as far away as Yorkshire, the Upper Thames Valley and the south coast of England. Ipswich's early prosperity relied on its proximity to the Orwell Estuary and thence across to mainland Europe. But many of the goods arriving in Ipswich for onward dispersal, and cargoes leaving Suffolk for distant parts, needed to go and come overland.

A little downstream of Stoke was the original point at which the river could be forded and later bridged. Known as Botford it remained an alternative route up to the late 18th century. It is thought that the ancient approach to the town from London was along Great Whip Street which led to the ford.

Most towns and cities start life beside a river and Ipswich is a prime example: its advantageous port status would have been severely curbed had the River Orwell not been bridged at several points very early on, long before the Conquest. But, what evidence there might have been of these early crossings was destroyed when work began on the Wet Dock in 1839. The traditional and ancient quays, and the course of the river itself, were altered out of all recognition. There may also have been a separate crossing named Peter's Bridge, known to have been maintained partially by the Borough, which also went by the name Ipswich Bridge but its precise location is lost.

In 1066 everything changed: the Normans succeeded in invading and colonising Britain and, among other things, began building moated castles. Moats by their nature required bridges and at **Framlingham** in 1101 Roger Bigod erected a motte and bailey castle surrounded by a ditch (although there is no evidence it was ever full with water). This was destroyed in 1173 but in the 1180s another of the Bigod dynasty regained the Framlingham estates and rebuilt the castle in stone. The castle changed hands many times in the ensuing years but one of its many features are the 13 square or rectangular towers along the 'curtain' wall; between each tower there would have been a wooden bridge that could be kicked away so that intruders were trapped with no means of entering the castle interior. Those on the Wall Walk today can still see where the stone pathway stops and wooden bridges take over.

There must have been several generations of wooden drawbridges at Framlingham Castle but between 1524 and 1547 a permanent bridge was built across the moat.

The use of drawbridges seems not to have been confined to moats as Mights Bridge on the road into **Southwold** was recorded as also being the site of a former drawbridge over Buss Creek. A bridge at this point was first recorded early in the 13th century and there had undoubtedly been a fording place long before then. What is now the A1095 has always been the only road into the town and as Southwold is a virtual island, with the sea on two sides, it would make sense for the town's defences to be at that point. The bridge is also known as the Buss Bridge and takes its name from the Herring Buss, a type of Dutch fishing-cum-cargo boat that used to moor there in the 18th century.

Eye, too, was a virtual island and the town grew up around a crossing of a tributary of the River Dove. The very name Eye is a Saxon word for island, the fords and bridges were therefore imperative to its survival and growth. There were numerous causeways out of the settlement over the ages of habitation: people have lived in Eye since the Late Bronze Age, a Roman hoard has been found there, and the Normans built a castle there soon after the Conquest. The Dove was crossed by Bolser, Abbey and Kings bridges, the river from Yaxley by Lambseth Bridge, and the watercourse through the Moor by Magdalen or Spittal Bridge.

It would be easy, but possibly erroneous, to point to **Woodbridge** as being somewhere that a name still preserves an early tradition of an ancient wooden bridge: however, although some chroniclers have chosen the obvious interpretation, the more general belief is that it comes from the early Saxon 'Woden' and 'burg', meaning 'Woden's town'. In the Domesday Survey of 1086 it appears as 'Wudebrige' and can, therefore, be interpreted as deriving from the Old English 'wudu' (wood) and 'brycg' (bridge). However, if there was a wooden bridge it is uncertain where it would have spanned the River Deben as the river is quite wide at that point and the first ancient crossing is a little further upstream at Wilford.

The modern crest of the Woodbridge Town Council contains the stump of an oak tree that refers to the wooden bridge from which the town is said to derive its name.

While **Bridge Street** is one of the surprisingly few surviving place name to incorporate the word 'bridge' there was, at Domesday, an area called **Bridge** near

Dunwich. It was held by the Norman, Roger Bigod, and consisted of '2 mills, saltpan, 20 goats' and had 7 households.

Eastbridge in the parish of Theberton is another 'bridge' name and its crossing stood on the ancient route from Blythburgh to Snape.

Also at Domesday was the Hundred of **Risbridge** (*Risebruge*) in the west of the county that contained 35 places and was converted to a Registration District in 1837.

Domesday also records that the street near the bridge in **Witnesham**, where the River Fynn rises, was called 'Fynford' and early medieval wills mention a bridge here, which would indicate that the ford was used sufficiently to warrant a permanent and more convenient method of crossing the river.

Stoke Bridge in **Ipswich** also finds a place in the Conqueror's Survey. The hamlet of Stoke was from an early period celebrated for its mills and in 1086 the then mill belonged to the monastery of St Etheldred. After the town was granted its charter in 1200 the town corporation also had mills either side of the bridge and it was compulsory for inhabitants to bring their corn to the town mills to be ground. The penalty for not doing so was a fine and forfeiture of a portion of the grain.

Just as churches were only listed in the Domesday Book if they had taxable assets, so bridges found little place in the Survey. Had they been of value as a source of income bridge history in Suffolk would have been carefully documented and read very differently.

CHAPTER 2

During the 13th and 14th century Suffolk was increasingly in the forefront of escalating trade and travel beyond the county borders. Fords and bridges were essential to the commercial success of medieval Suffolk both for the transport of goods, livestock and people, including those attending fairs, markets and places of pilgrimage.

Travel overland was difficult enough, especially on the heavy claylands, and improvement in river crossings was both inevitable and vital. Where logs gave way to basic wooden bridges they were, in due course, replaced by worked timber, then stone and/or bricks, until it became possible for pack animals as well as people to pass over rivers and creeks. But it was not until the 13th century that wooden bridges began to replace ford crossings in any number and invariably the fords were kept alongside. Wooden bridges were often makeshift constructions that could not take much weight and were prone to being swept away by storms or tidal surges.

At some point it was realised that a straight 'beam' bridge did not distribute the weight evenly and was liable to snap if too much was put on it. An arched bridge was the solution but it took time and money to build, so they were slow to catch on in rural areas

Many other parishes which have fords or low-lying roads have raised footpaths, usually consisting of a series of wooden planks or, as at **Bruisyard,** a narrow but solid gangway with stout railings either side. At **Kettleburgh** the raised planks start a few hundred yards from the bridge and there is a similar arrangement not far away at **Glevering**. On the edge of **Ash Street** is a watersplash and Irish bridge.

The Grade II listed red brick bridge at **Glevering** spans the River Deben but only half of it lay in the parish, the other half is in **Hacheston.** It is dated 1777 (on the keystone of its main arch) and probably replaced an earlier wooden crossing.

At **Stoke Ash** in Deadman's Lane just east of the A140 the bridge and a raised ford crossing run parallel to one another. This raised footpath is invariably referred to as an Irish bridge as is the one at Bruisyard, although there is no clear definition of what constitutes an Irish bridge. **Finningham** has an Irish bridge on Ladywell Road and there are raised walkways at **Haughley** and **Onehouse** and many more besides.

Bridge building played a large part in making early 16th century Suffolk the fourth richest county in England as well as the most densely populated. As to be expected these new bridges did not appear either overnight or by magic: they had to be paid for and thereafter maintained. In fact, who did the paying had been a subject for discussion and disagreement from the earliest of times. Many private individuals contributed, as did the relevant borough or local authority, and the Established Church became a main player. The influence and involvement of the Church in the daily life of the population grew steadily from the 10th century onwards and the increasing numbers of priors and abbots were among the first to charge people to cross their bridges.

The pressure on landowners and civic authorities to provide improved crossings was sufficient for bridges to be mentioned in 1215 where Clause 23 of the Magna Carta states that 'no village or individual shall be compelled to make bridges at river banks, except those with an ancient obligation to do so.' Itinerant justices and sheriffs were required to apportion responsibility for bridges but although they were at liberty to ascertain the condition of the structures they had to rely on local information as to liability of repair costs, which was not always forthcoming. No one wanted to pay for a bridge unless they were legally obligated so to do and once the sheriff had moved on, little was achieved.

From at least the 8th century landowners were obliged to provide what has become known as the 'common burdens', i.e. providing military service, labour for the building and repairing of bridges and workmen to build and maintain fortifications. However, while the spirit of the common burdens prevailed in Suffolk they are not recorded in any Anglo Saxon charter. The reason would seem to be that the East Anglian kings ruled independently but were well aware that bridge work was considered to be a fundamental aspect of travel across Suffolk. They needed tried and tested crossing places across rivers, as did their subjects, from whom they demanded the same degree of public assistance as elsewhere in the country but they were not bound by the same rules as those outside their kingdom.

Gradually, though, the common burdens receded into history and by the 11th and 12th century whenever a bridge collapsed or was swept away there was invariably no one who was legally or morally obliged to accept financial responsibility. A local Lord of the Manor or religious institute might step in, for reasons of their own, but the physical repair of bridges was not a legal obligation on any one person or public body.

When it came to bridges there were three main choices: lordly or royal inclination, charitable or religious impetus, or community action by way of a bridge trust or similar. The obvious way of recouping repair or renewal costs was to extract a toll from those using the bridge.

One of the earliest documented tolls was levied on the 12th century at **Sudbury** where there was an ancient crossing over the River Stour likely to have been used from at least Roman times. Known today as the Ballingdon Bridge, the early crossing was built by William Fitzrobert, Earl of Gloucester, and the tolls were collected by staff from the hospital that once stood on the site now occupied by The Boathouse.

Sometime later the Ballingdon Bridge tolls were in the hands of Amicia, Countess of Clare. At the start of the 13th century tolls were collected on behalf of the Countess from those using the bridge that had become indispensable to travellers on what was an important ancient route from Braintree in Essex to Sudbury and Bury St Edmunds.

The bridge at **Brandon** was also a toll bridge and its traffic was pilgrims on their way to the Shrine of Our Lady at Walsingham in Norfolk. The Little Ouse was too deep for a ford so before the first bridge was built people and goods were conveyed across the river by ferry, hence the still-used name of Brandonferry. There is much discussion to be had as to the exact location of Brandon's original bridges, namely Fengate Bridge and the West Bridge, but it is generally agreed that the bridges, wherever they were, brought prosperity to Brandon.

The importance of bridges in country areas during mediaeval times is illustrated by Robert Reynes of Acle in Norfolk in his *Commonplace Book*, written in the last three decades of the 15th century. He sets out to count the miles from Acle in Norfolk to Canterbury in Kent not from town to town but by measuring from bridge to bridge thus emphasising their importance. When he reached East Suffolk he jotted down, '10 miles to Beccles, 7 mile to Blythburgh, 4 mile to East Bridge, 5 mile to Snape, 7 miles to Woodbridge, 5 mile to Ipswich'.

The earliest record of **Blythburgh** bridge over the River Blyth is found in a priory charter of about 1200, where smaller footbridges are also cited. A 'great bridge' over the Blyth is mentioned in 1296 and is thought to have stood near the site of the present modern one. It would have been the Prior's responsibility to repair and maintain the bridge until the Dissolution in 1537 and reflects the influence of the Established Church in such matters.

In medieval times it appears there were two or more other bridges in Blythburgh: a track from the church to the river leads to the site of an old bridge identified as 'Campys' bridges. Blythburgh Historian, Alan Mackley, points to a 20th century postcard showing a cart in shallow water at that point but while it may have been fording the river it could also have simply been soaking the cart wheels to tighten the metal tyres.

One of the first wooden bridges at **Snape** over the River Alde was almost certainly built by the Benedictine monks from the nearby priory but over the years there were numerous disputes as to who should pay for its upkeep. In 1347 the Earl Marshal was asked to repair the bridge as he held lands on either side of the river and by 1492 it was in a poor state and in need of substantial repair. The Bishop of Norwich gave permission for 'alms' to be sought from travellers using the crossing to defray the costs so initiating the first official tolls at Snape Bridge.

At **Sibton** there is evidence beneath the present bridge for earlier stonework from a structure built in the 12th or 13th century by the monks of the Cistercian Abbey over the River Yox. It was rebuilt in 1770 by John Freston Scrivener and still stands on the south-west corner of the abbey ruins. It has a semi-circular single arch and although there is a keystone it is impossible to read what was once written there. Sibton Abbey was the only Cistercian house in Suffolk and established in 1150 by William de Cadomo, a descendant of the powerful William Malet, tenant in chief of the manor of Sibton at the time of the Domesday survey.

The Sanctuary Bridge at **Letheringham** was close to a 14th century Augustinian priory and its name would indicate that here or nearby was a place of sanctuary for someone pursued by enemies or the law. Perhaps there was a cross there similar to the stone one discovered at Great Ashfield bridge which was thought to mark the sanctuary for criminals (see Chapter 3).

A thief seeking sanctuary at the Priory in **Clare** found the concept of protection in a holy place to be wholly unfounded. In 1385, Sir Thomas Mortimer and a posse of knights seized the thief from the Priory Church and cut off his ears. The Bishop of London, on hearing that the man's sanctuary had been violated, ordered Sir Thomas and his men to do penance by walking barefoot and bareheaded down Mill Lane and over the bridge to the Priory. They were clothed in rough garments and carried heavy candles. They were required to lay the offerings of the candles, together with a rich cloth of gold, on the altar in order to be absolved from their sin.

Abbots Bridge in **Bury St Edmunds** is one of the oldest surviving bridges in Suffolk; certainly it is one of the most picturesque. This 12th century mural (walled) bridge spans the River Lark and runs from Eastgate Street across to the opposite bank where, in the pre-Reformation days of the great Benedictine Abbey, the vineyard grew. It was constructed either during the abbacy of Anselm, who died in 1148, or that of Sampson, who died in 1211. The bridge gives passage to the flow of the river and on the north aspect can be seen the three arches along which a footbridge was created by laying wooden planks from buttress to buttress, around 5 feet wide. There is another footbridge on the other side of the wall which is plain and has no arches.

In the 18th century there was still a footbridge and ford just a little way down from the Abbot's Bridge which was used by the townspeople, and their horses, the river flowing down the street.

Early medieval Bury St Edmunds had five gates to the town and when Southgate Street was widened in 1970, road works revealed a medieval bridge spanning the Linnet. It closely resembled the Abbot's Bridge leading to speculation that there were other such bridges built in the 12th century but now lost.

Bury St Edmunds was dominated by the abbey and its bridges were largely dictated by the comings and goings of those visiting it, together with the usual traffic connected with the town's markets and fairs. Other places on commercial routes also owed their improving road and bridge structures to increased prosperity.

The very busy medieval bridge at **Hadleigh** over the River Brett was an important component in routes connected with the wool trade and thus became an early toll bridge. Flemish weavers arrived in Hadleigh in the 13th century and made it a thriving centre for the wool and cloth trade. Known as the Toppesfield Bridge it crosses the river south of the town and is the oldest bridge in the county still carrying vehicles. Although it is said to have been built in the 14th century it looks as though brick and stone of an earlier date was used in its construction. The bridge has three slightly pointed arches with a low parapet each side. The volume of traffic caused it to be widened in the 16th century

As might be expected fords in the countryside took much longer to replace and were invariably subject to local disputes as to who should do the paying. Those on trade routes to London and elsewhere were built and maintained with more enthusiasm as were bridges popular enough to yield tolls. Those in the towns, particularly Ipswich, found sponsors much more easily.

The **Stoke Crossing** over the River Orwell was certainly in existence by the 13th century as witnessed by the 1301 will of Thomas Aylred in which he assigned an annual rent towards the maintenance of the bridge to be administered by the bridge custodian. The presence of a custodian illustrates the importance of the bridge and it must surely have been one of the earliest in the county to extract tolls. Thomas himself benefited from having a well-maintained bridge as he not only held property in Stoke and Ipswich but was a prominent businessman and an officer of the Borough of Ipswich.

No doubt the fluctuations in the fortunes of Ipswich over the years would account for the condition of the Ipswich river crossings: when the town was doing well men like Thomas Aylred were willing to spend money on good access routes. The 16th century was a prosperous time while in the 18th century the town suffered a severe depression when the cloth trade declined.

Even after a bridge was thrown over the river (a timber bridge is known to have existed in the 1470s and a stone one in 1669) the Stoke ford and bridge operated alongside one another. In the 15th century cart wheels 'shod with iron' began to damage the paved streets and bridges. Horses and wheeled traffic were often denied the use of the bridge, directed instead to enter the town by way of the ford. Indeed, in 1477 the bailiffs were instructed to keep the bridge locked against carts and a few years later decreed that 'none shall go over the bridge when they may go through the ford'. The two were still operating in tandem in the 19th century. During 1804 considerable repairs to the bridge meant that all vehicular traffic had to be diverted 'to the Ford at St Peter's Dock'.

It is not always clear in old documents which bridge is being referred to but in the 1430s a John Caldwell offered to repair 'the bridge' at his own expense if the borough would contribute to its maintenance. This could be Peter's Bridge, Ipswich Bridge or Port Bridge as all three appear at one time or another in the annals. By the 15th century another was called North Bridge while that closest to Stoke is called Stoke Bridge. The entire crossings configuration was lost in the years 1839 to 1842 when the Wet Dock was constructed.

The second most important river crossing to aid the prosperity of **Ipswich** is the Handford Bridge, which is further up-river from the Stoke crossing but is certainly as old. The ford must have existed for as long as man has inhabited that part of Suffolk and was used to approach the town by skirting the northern edge of the medieval marshes. It is described on John Ogilvy's 1674 map of Ipswich as 'the Handford Bridgeway' although, unlike Stoke, it seems not to have had a gate or bridge custodian.

Another historic bridge in **Ipswich** survives in name only in one of the town's smallest thoroughfares, Friars Bridge Road. In ancient times there was a bridge and causeway over a tributary of the Gipping towards water meadows and marshlands called Oldenholme. However, shortly after King John granted Ipswich its charter in 1200 the town ramparts were built, leaving the bridge and meadows outside the town defences. When the pre-Reformation Franciscan Friary was established thereabouts in 1290 it became the Friars Bridge and marked the western boundary of the Friary which extended south from Friars Street. The bridge was also used by the town portmen who were given pasturage for their horses on the meadowlands. A clue as to how the single-arched humped back bridge it looked in the 19th century can be seen in Henry Davy's watercolour of Friars Bridge, executed in 1837, on which he wrote 'this stream joins the Gipping.'

On Hodskinson's 1783 Plan of the Town of Ipswich, Friars Bridge (spelled Friers) is clearly marked, leading westwards towards Portman Walks, a short way down-river from the Handford crossing.

Yet another **Ipswich** bridge once stood where Major's Corner now is but in 1659 the Town Surveyors agreed that the bridge could be removed so that 'the Water may Runn downe the Lane'.

Construction of decent bridges elsewhere over the old fording places greatly improved journeys, particularly in the 15th century when more farmers and tradesman began using horse and carts to attend markets.

At about this time myths and legends about bridge dragons began life, one of them at **Bures** in the Stour valley. In 1405 a fire-breathing dragon took up residence by the nearby bridge, 'a dragon vast in body with a crested head, teeth like a saw and a tail extending to enormous length'. It killed a shepherd and his sheep before the village men attacked it with arrows, which did it no harm as they only bounced off. But the dragon took the hint and one day swam off downstream and was never seen again.

At **Little Cornard** in 1449 there were two dragons, a spotted red one from Ballingdon Hill on the Essex side of the Stour, and a black one from the Suffolk side. The two dragons met on the bridge, no longer there, and did battle. Finally the red dragon won but both creatures survived the duel and each returned to their own lair, never to be seen again.

The importance of these crossing places in the everyday life of medieval society is illustrated by the numerous bequests in 15th century will for the reparation or maintenance of local bridges and roads leading to and from them. In medieval Christianity the building and maintenance of bridges was akin to a pious act.

In 1463 Agnes Jenews of **Eye** left 6 shillings and 8 pence 'to the emendation of Lamesheth bridge'. Her husband John made a similar bequest in 1448/49 'to the emending of Lamesheth bridge 6s 8d'.

In 1477 Agnes Dyke of **Stoke by Clare**, widow, left 13 shillings and 4 pence for the reparation of the bridges at 'Babthernbregge' and the way 'between the said bridge there, on the riverbank.'

At **Haverhill** in 1474 Thomas Lefeld the Elder left 6 shillings and 8 pence to the reparation of the bridge called 'Belyngesbregge'.

The bridge at **Beccles** was an important and much-used spanning of the River Waveney and led travellers out of Suffolk into Norfolk and *vice versa*. At last count there were over a hundred Beccles wills that mention the bridge but one of the earliest is from 1388 when Edmund de Wellis, Rector of the parish, left 6 shillings and 8 pence for the bridge. John Monk left one shilling 'for the foundations of the bridge' as did Edmund Candeler who left 6 shillings and 8 pence 'for the bridge and causeway'. From these few extracts it would seem that work was being carried out on the bridge regularly and indicates its crucial role in town business.

The **Brandon** ferry and bridges constituted another important crossing between Suffolk and Norfolk and was referred to as the 'thoroughfare town'. In the 13th century a wooden bridge was built and local tenants, called Briggemen, were made responsible for its upkeep and maintenance. In the early 14th century the Bishop of Ely commissioned a stone bridge to replace the wooden one at around the time permission was granted for a market and town fairs, which attracted visitors from near and far and took trade away from its rival, Thetford (in Suffolk until 1889). Brandon, one of the few Suffolk bridges where permission to charge pontage (a toll levied at a bridge for its maintenance) was requested and granted in 1330 for five years.

Anyone writing about Brandon before 1953, when the bridge was demolished and re-built, eulogised about the historic elegance and charm of the ancient structure. In 1948 Olive Cook wrote:

> A five-arched bridge of medieval irregularity crosses the river, the natural stronghold of idlers and a coign of vantage from which to watch the swifts swooping and crying about the housetops, and to follow the course of the slow stream through a landscape of sandhills and marsh where starveling alders and melancholy pines stand knee-deep in water.

In 1942 Doreen Wallace wrote of Brandon's considerable beauty enhanced by 'a stout and ancient single-track bridge over the Little Ouse, one of those bridges with refuges for pedestrians at intervals.'

The reconstructed bridge does still have two pedestrian refuges on each side and is only one of three Suffolk bridges to have them (the other two being Wilford and Bourne Bridge in Ipswich).

Brandon Bridge featured in many guide books and was known throughout the country for its individual character and charm. Thousands of pilgrims passed through Brandon in pre-Reformation times on their way to the Shrine of Our Lady at Walsingham, and many a craftsman or stonemason must have crossed over the bridge on his way to the flint knapping yards for which Brandon was famous.

In the 13th century, in the reign of Edward I, the River Waveney could be crossed at **Herringfleet** by means of a ferry and operated approximately on the site of the modern St Olaves Bridge. A boat was used to convey passengers from St Olaves Priory across the river between Suffolk and Norfolk. Its ferrymen had strong local support although its operation led, over the years, to considerable disagreement between the two counties.

An inquisition was held in 1296 to enquire 'what detriment it would be to anyone if leave were granted to Jefery Pollerin of Yarmouth to build a bridge at St Olave's Priory.' The verdict was that while it would indeed be detrimental both to the holder of the ferry and to the Prior, a bridge would be of great benefit to the country. No bridge was built and the ferry continued unopposed. Permission was given again in 1420 for a bridge to be built but no bridge materialised.

During the 13th century the ferry was kept by a fisherman named Sireck who received 'bread, herrings and such like things' to the value of twenty shillings a year. When he died his son William became ferryman and was paid ten shillings more than his father. William's two sons Ralph and John both, in their turn, followed in the family tradition.

Late in the 14th century John de Loudham became Lord of the Manor of Herringfleet and the ferry became the subject of a petition to Chancery. In around 1400 John de Loudham claimed that he held the ferry through a grant of John ate Ferye whose family had held it 'from time without memory'. The arguments raged back and forth for some years.

Eventually, though, a bridge became a necessity. Some time towards the end of the 15th century a three-arched stone bridge with four turrets was built on the instructions of Dame Margaret Hobart, wife of Sir James Hobart, Attorney General during the reign of Henry VII. A painting in Loddon church (Norfolk) show a three-arched bridge with two horses and two drovers in the corner above Lady Margaret (other copies show four arches). The inscription includes 'pray for the soul of Dame Hobart, wife of the aforementioned James, who built the Bridge of St Olaves with a hard core highway leading to it at her own great expense, for the public good.'

It was called the Hobart Bridge and its construction finally ended years of acrimonious argument.

Perhaps the best known of all Suffolk's early bridges is that at **Moulton** often referred to as a packhorse bridge and famous the world over. It was built around 1446 and spans the River Kennett on the old route from Bury St Edmunds to Cambridge. But the river has shrunk in volume since the four-arched bridge was built and it stands parallel with a concrete ford that takes traffic alongside the bridge. The cutwaters are plainly visible and give an indication of the quantity of water that it had to contend with in former days. Its original cost would have been considerable as would the cost of maintaining it: to help with repair expenses a charity known as 'The Church and Bridges Estate' was set up by the parish in around 1530 and rents from a small estate in Freckenham, together with 13 acres around Moulton, provided an income for both the bridge and the church.

The length of the bridge way provided a gentler slope than a single span construction which would facilitate the loaded ponies, especially in winter when the surface would be slippery.

Most of the packhorse trains had anywhere between ten and fifty horses. Reporting on the Suffolk Institute members' visit to Moulton in 1932, Arthur A Watkins wrote in the Proceedings:

> The horses walked in single file, attached to one another. The halter of one horse being tied to the tail of the horse in front of him. The leading horse, as it bore a bell, or collar of bells, to call attention of approach, was called the Bell Horse.

All commodities, such as corn, wool, cloth, poultry, fish, barrels of butter, salt, hops and household articles were transported to market on the back of horses and even manure was carried in panniers to the fields and peat and wood for winter fuel brought to the farmhouse in the same way. However, while these packhorse bridges may well have conveyed pedestrians, small carts and possibly single horses or ponies over the rivers, larger carts and long trains would have used the original ford crossing. Perhaps if they were carrying silk, spices or anything that could be damaged by water, an individual horse and driver would use the bridge rather than risk the ford. The bridge would anyway not have been able to accommodate heavier and larger loads and the horses could often perfectly well cross the fords which ran alongside.

The Moulton bridge is constructed of flint and stone rubble but the edges of the arches are brick. The pointed shape of the arches was created by first building a timber framework and laying the bricks in place on these 'formers', which were then removed. These pointed arches are typical of 13th and 14th century bridges, the design superseded by bridges which were ribbed under the arches while others had semi-circular arches.

There is a flint and stone footbridge of similar design nearby, of uncertain date, which has a long single arch and centrally raised walkway.

Although Moulton is referred to as a packhorse bridge, that is horses loaded with panniers or side bags, it was probably built originally to accommodate the quite small ponies and carts that had been used since time immemorial. Local farmers usually travelled between hay making and the corn harvest when the roads were drier and they could reach the markets with relative ease but no doubt they took advantage of the bridge when they could at other times.

It has always been thought that the parapets were kept low so that the panniers would clear them but modern thinking is that these low sides were not necessarily designed for panniers but to keep construction costs down. The walls of the bridge needed to be just low enough to stop the cart wheels going over the edge and no higher. However, this bridge is always known as a packhorse bridge and will no doubt continue to be so.

The packhorse bridge at **Kentford** was contemporary with that at Moulton and sometimes known as the Old Roman Bridge. It had four arches of flint rubble and brickwork that were still visible up to the mid-1970s when they were washed away in floods.

It is not clear when it was first built but Walter Kyngesman, a glover of Bury St Edmunds, died in 1462 and left 'to the sustentation and emendation of the stone bridge in the town of Kentford, for my soul, 12d.'

In 1741 an anonymous author wrote *Journal of a Tour through Suffolk, Norfolk, Lincolnshire and Yorkshire in the summer of 1741* (now in the British Library) and said of Kentford:

> Over the brook is a very high bridge for horses. The brook, upon suddain rains, being often impassable. On each side of the bridge is a wall built very compactly with flints which is now kept up at the charge of the parish and this is so old that the inhabitants have no reports as to the foundation of it.

The remains of another packhorse bridge, known as Stone Bridge, are at **Cavenham** where it crosses a tributary of the River Lark just west of the southern section of ancient defences, Black Ditches. Described as being 15th or 16th century, it was said to be only 12 feet wide, with a single span of four courses, and no parapets. It stood on the Icknield Way, across a small tributary of the River Lark, and its size would barely accommodate a single horse or two people abreast at any one time. If there were ever parapets the width would have been further reduced. It was re-bricked in the 18th century and it is difficult to know what alterations were made at the time.

There is a pumping station south of Stone Bridge and the nearby Robert's Bridge which accounts for there frequently being no water under either bridge.

This ancient crossing is now barely discernible and only protected from traffic by an insubstantial iron rail, but it is, nonetheless, only the second such bridge still in some kind of existence, the third if you count the one lately at Kentford.

There is reputed to have been a packhorse bridge near **Risby** which is not unlikely as Risby stands between Cavenham and Bury St Edmunds. The bridge would need to be somewhat west of Risby itself as the village does not straddle the stream.

At **Thetford**, on the confluence of the rivers Thet and Little Ouse was situated an ancient fording place where a 16th century packhorse bridge once stood. This was historically an important crossing place and takes its name from the Anglo Saxon 'Theod' (ford), 'the people's' or 'main' ford. The River Thet took its name from the town.

The present Thetford bridge, known as Town Bridge, was built in 1829. It was previously known as the Christopher Bridge because the Christopher Inn stood close by. Previous bridges at that site were toll bridges, which provided a sizeable income for the town in the 16th to late 19th centuries.

The Town Bridge replaced a wooden structure built in 1797.

Hempyard Bridge at **Ixworth** is invariably referred to as a packhorse bridge possibly because it was thought to be on a drovers' route bringing cattle down from Scotland to the London markets in the 1800s. The bridge, currently in a bad state of repair, probably began life as a crossing for the nearby Augustinian Abbey and is one of those bridges where people can linger and ponder its history without interference from road traffic.

Towards the end of the mediaeval period roads and bridges had become an accepted and indispensable means of transport, making inland travel less dependent on boats. Communities grew and became established at crossing points. Inns and taverns flourished on either side of the bridges. If the watercourse was tidal thirsty and hungry travellers needed somewhere to wait while high tide receded. Gradually the unbridged routes gave way to the increasingly popular bridged ones.

Many of the more ancient routes from time out of mind lapsed over the centuries and are thus unrecorded and lost for ever. A few do survive in place-names although the old fording places are not necessarily in the same location as the modern bridges. It does not take long for a disused causeway to become overgrown and as towns and villages expanded or sometimes waned so, too, did the fords and roads that linked them.

By 1500 practically all the now known watercourse crossings were in regular use and most significant fords were bridged. From the start of the 16th century until the 1750s few additional bridges were added to the total. The ensuing routes were, by and large, those which became the basis for the on-going road system based on trading and general travel and were in due course turnpiked. Gradually the most-used roads became the main routes for almost two hundred years until the age of first the train and then the motor car when the road system altered dramatically for the first time in history.

One of Ufford's curved parapets

First of Chelsworth's two bridges

Sarsen stone beside Boxford Bridge

Raised footbridge leading to Kettleburgh Bridge

Sibton Abbey ruins and bridge

Knettishall, one of Suffolk's prettiest bridges

Almost high tide at Wilford Bridge

Hoxne's famous Goldbrook Bridge

Detail from painting of St Olave's bridge in Loddon Church

Abbot's Bridge in Bury St Edmunds

CHAPTER 3

The word hermit conjures up images of a solitary religious recluse but a bridge hermit could also be a public servant, a guide, ferryman or even a river pilot and in practically all cases he would be appointed by the bishop. From the 6th century through to the Reformation, bridge hermits performed a variety of duties and were 'encouraged' to seek alms for the bridge upkeep. Some were more gregarious than others but their role was primarily looking after the fabric of the bridge or causeway (and the bridge chapel, if there was one), minister to those who used it and be responsible for collecting tolls if applicable.

At **Cattawade,** a hamlet of **Brantham,** the bridge hermit was responsible for the bridges, at one time there were more than one, and the bridge chapel built in honour of the Blessed Virgin Mary. In 1350 Roger de Kenton, warden of the bridge, applied for a grant of land 'on the high road, to himself and his successors' in order that he might build a chapel by the bridge.

Jervoise records that confirmation of the grant is recorded in the Patent Rolls of June that year:

> The plot of land was to be 100 feet long and 48 feet broad, parcel of the King's highway by the causey of the bridge.

Ten years later, in 1360, the hermit of 'Cattiwade' was granted permission to seek alms towards the maintenance of the bridge and causeway.

In 1359 the hermit John atte Welle, together with John Canes of Brantham, were keepers of the bridge and causeway between 'Manytre and Cattiwade', that is, Manningtree (Essex) and Cattawade, over the River Stour. Travellers passing from Suffolk into Essex were afforded the ministrations of the keepers and an opportunity to pray at the chapel for a safe journey or give thanks for a safe return.

Cattawade was granted a market in 1247 and good access was a must, as it was to the other medieval markets across Suffolk. In 1256 Walter de Suffield, Bishop of Norwich, left the sum of 2 marks towards the repair of the bridge indicating its importance in the local economy.

It is to be wondered if any of these keepers of the Cattawade crossing came from the nearby Dodnash Priory, founded in the 12th century. Those in the priory would certainly have used such a bridge. A single stone is all that remains of the priory: it is about a foot high and stands just north of the stream which serves as the parish boundary between **East Bergholt** and **Bentley**, to the south of Dodnash wood. The bridge on the Bentley to Brantham road passed by the priory and a monk is said to haunt the causeway, though no one has ever said why.

There is also a tradition that says treasure is concealed beneath the stone although if such was the case it would surely have been robbed a long time ago. What the treasure could be is anyone's guess since the small priory itself had no great endowment and was barely viable throughout most of its existence. Dodnash was one of those priories dissolved in the 1520s by Cardinal Wolsey seeking funds for his new Oxford college.

The 16th century farmer-poet Thomas Tusser lived for a while in Cattawade and would, no doubt, have crossed its bridge on many occasions. He was a restless soul and did not stay very long in one place. It was while he lived in Cattawade that he wrote *Five Hundred Points of Good Husbandry* a collection of hints and advice for farmers written in verse.

Another legacy of the ambitious plans of Cardinal Wolsey for his Ipswich College is **Smallbridge Holt** on the River Stour at the Suffolk-Essex border close to **Bures St Mary**. The Holt, or Wood, is at Wormingford Bridge and in 1523 Wolsey began to fall foul of his royal master, Henry VIII, and those lands he held there fell into the hands of the Waldegrave family. At the time the bridge was named Small Brigg but Sir William Waldegrave converted the Essex side of the river into a deer park and connected it to his grand house by a new bridge and changing its name from Small Brigg to Smallbridge.

At the bridge in **Semer** stood the chapel of the Blessed Mary of Neddynge (Nedging) and in 1472 Ralph Smyth of Bildeston left a bequest to the chapel of '4 bushels of barley and 3s 4d.'

Semer Bridge was one of those mentioned in the bequest of cloth merchant John Motte of London who died in 1473. Motte, who based his operation in **Bildeston** had a team of spinners, fullers and weavers working for him: when he died he not only remembered his workers but left the huge sum of £240 to maintain roads and bridges including those at Bildeston, Semer, Rattlesden and Ipswich. His investment in the transport system of rural Suffolk was one factor that made him, at one time, the fifth largest clothier in the county. It illustrates the relationship between good infrastructure and prosperity: men like John Motte could not have functioned without decent bridges.

Sometimes there was a chain across a bridge, the hermit holding the only key. The toll went towards maintaining the structure and the causeway leading to it. At **Beccles** though, the hermit of the Great Bridge across the River Waveney not only had to look after the bridge but also had responsibility for the bridge chapel of the Blessed Virgin Mary as well as taking part in the general life of the community.

In 1455 the name of the Beccles hermit was William Warde and he is named as executor of a will. A few years later, in 1462, he is charged once more to carry out the 'reparation of Beccles Bridge.' In 1463 Warde is again named in connection with 'the reparation of the chapel of the Blessed Mary at the Great Bridge of Beccles.'

At **Brandon** the hermit was empowered to carry out priestly duties in the chapel of St Mary and St Etheldreda which stood on the bridge over the Little Ouse river. For centuries the river had been crossed by ferry but in the 1240s the Bishop of Ely decided it was time for a wooden bridge. To offset the cost, tolls were charged for travellers using the bridge, a wooden one until 1330 when it was replaced by stone and pontage for a period of five years was granted to the Bishop for 'the repair of the bridge at Brandonferry.'

Not all bridge hermits get mentioned in despatches but there are names for some of those at Brandon: 1406 Hermit William Bussheby was appointed, followed by John Newton, Richard Passhlew and Thomas Passhelawe (who died in 1459). In 1459 William, Bishop of Ely, instructed his bailiff John Pepper to install Hermit John Herryman in the Brandon Bridge hermitage to replace Thomas Passhelawe.

Later, in the 1550s, the right to impose tolls was lost to the church and granted instead to Sir Richard Fulmerston by Edward VI for a fee of £4 a year. In 1573 the town of Thetford gained the right but with it went responsibility to 'repair, sustain and amend' the Brandon bridge. (Thetford itself, with its many bridges, had sufficient toll income to warrant a Toll House which was not demolished until 1880 long after tolls had been abolished.)

Much of the Brandon bridge traffic, and therefore income, came from pilgrims on their way to the Marian shrine at Walsingham (Norfolk), then the second most important shrine in England after Canterbury.

In order to persuade people to contribute towards the maintenance of the chapel, bridge and causeways the bishop directed the Brandon hermit to offer 40 days 'indulgence' to those who contributed. An indulgence was granted to a Christian confessing his or her sins after which the priest could give various forms of absolution from guilt. Indulgences could be 'bought' by good deeds or by donating money for the use of the Church or, in this case, for the repair of the bridge chapel. At Brandon the 40 days indulgence was offered at regular intervals – 1459, 1475 and in 1497, for example, when repairs were needed to 'lights, ornaments and rods'.

To augment the tolls and indulgences the Bishop also instructed a fishery to be set up, the proceeds of which would go towards the repair of the bridge.

The river crossing at **Cavenham Heath**, south west of **Icklingham,** straddles the River Lark on its way to the Great Ouse and is known as the Temple Bridge. There was a crossing here at early as 1180 and during early medieval times it reaped a toll income for its ecclesiastical landlords. It is north of the packhorse bridge at Cavenham and it is possible that this crossing was an alternative route for London travellers on their way to Thetford and Norwich. Both Cavenham and Icklingham were on the old Pilgrim Path that eventually met up with the Icknield Way.

Naturally enough, Temple Bridge has been rebuilt and replaced many times since the 12th century and several legends have grown up around it. One concerns the ghost of a headless rider, lightly clad and riding a horse over the low meadows by the bridge. Some think this headless rider is a Roman soldier presumably because there was a Roman station at Icklingham. Others think it is more likely to be Sir John de Cavendish, Prior of the Abbey at Bury St Edmunds, who was pursued to the bridge by the mob during a period of violent town revolt against the abbey. Sir John was captured and beheaded in 1366. A third option is the headless ghost of the Archdeacon of Sudbury who is said to constantly walk the site of his murder on Temple Bridge.

At **Bungay** there was a very early bridge hermitage and a small adjoining chapel at the east of the bridge on the south bank of the River Waveney (between Bridge Street and Falcon Lane). It is believed that the bridge hermitage was founded in the 14th century but most of its fabric was demolished in 1733 and a granary built in its place. In 1798 a few large stones were found near the river, which were said to have formed part of the hermitage's north gable. There is a possibility that the Friars Minorities, or Franciscans, might have been associated with the hermitage: ancient walls at the entrance to Bridge Street are said to have belonged to a religious building and thus part of the bridge chapel complex.

Jervoise records that a bridge chapel to Our Lady was built at Bungay in 1532 although this must have either been secondary to the earlier one, or a replacement. However, if this date is correct it was completed only a few years prior to its being dissolved as part of the Reformation.

In 1352 the hermit's bridge and the bridge on the Bungay to Earsham (Norfolk) road, were both reported as having been 'overturned and submerged by heavy rains.' The Lord of the Manor and Castle of Bungay agreed to supply timber for the repairs but the townsmen must cart it from his wood and carry out the work at their own cost. Another 'great water flood' occurred in 1696 which destroyed the main Bungay Bridge and at least two others.

Bridge hermits were invariably called upon to execute will bequests to their chapels, which practice continued into the 16th century. In 1511 Alexander Rycherdson of Walberswick made a bequest to the Holy Rood Chapel at Blythburgh Bridge. Alfred Suckling wrote in 1846 that the chapel was erected on the north side of the main street in Blythburgh leading to the bridge and that some remains of the old walls were standing in 1754 when Thomas Gardner published his history.

In the 16th century **Hadleigh** had two chapels, one of which was attached to Hadleigh Bridge (north of Toppesfield Bridge) that spanned the River Brett. It was dedicated to the Blessed Virgin Mary and travellers would light a candle to implore for her intercession for a safe journey or in thanksgiving for a safe return. An old guide book says:

> Before crossing the bridge ... records dating back to the 15th century indicate that here stood a Chapel of Our Lady where returning travellers might pause on their entry to the town. We cannot say exactly where the chapel stood - it may even have been across the road on the town side of the river, where today there lies a piece of the iron bridge built in 1843 in place of the old medieval structure and replaced for safely reasons in 1979.

There is no mention of a hermit as such at Hadleigh but there would have been someone in charge of the chapel and the bridge who would have performed a similar role. In 1252 Hadleigh was granted a Monday market which ensured regular weekly traffic both to the market and back out to the villages and other clothing towns to the west.

There was probably a bridge hermit at one of the **Sudbury** bridges who received alms for the maintenance of the structure and kept the causey in good repair. Sudbury was one of the very few who applied for a grant of pontage which was granted in 1331 for three years.

The bridge chapel at **Stratford St Mary** was dedicated to St John and stood just south of the bridge. It was demolished in 1826.

Perhaps because of its name, one of the best known toll bridges is the Tu Penny Bridge at **Eastbridge.** Sometimes known as the Minsmere Toll Bridge its origins are obscure and no one is quite sure when the toll was imposed or by whom. Its name suggests that it cost two pennies to cross but whether or not it was a private bridge, or one under the jurisdiction of a local abbey, has never been fully established. It is marked as a toll bridge on Hodskinson's map of 1783 but it is presumed to be an ancient crossing and the toll could also be ancient. The nearby Eel's Foot Inn dates back to the 1640s and was no doubt patronised by many a 'tu penny' traveller.

The ancient Abbot's Bridge at **Bury St Edmunds** is one of the most complex bridges extant in Suffolk and the keepers of the five town gates, including this one, would have been monks from the Benedictine Abbey. Each of the gates housed a chapel to Our Lady in which a light was kept burning continuously. The East Gate immediately adjoined the Abbot's Bridge and guarded both the ford and footbridge and the keeper's cell can still be seen.

A second medieval bridge that spanned the Linnet was discovered in 1970 and seems to have closely resembled the Abbot's Bridge. There were also likely to have been several small bridges over the moat-like ditch that ran along the outside of the west and north side of the Abbey precincts.

Streams and rivulets abounded in **Eye** giving rise to numerous causeways and bridges but in accounts drawn up in 1458 by the two town chamberlains, a new bridge was built called 'le spetelbrydge' and there may well have been someone on the bridge seeking alms. The 'spetel' was the leper hospital of St Mary Magdalen which lay outside the town (in what is now Magdalen Street). The hospital had been there from the time of Edward III and founded sometime round 1329.

Will bequests were made for 'le spetel', for example, in the early 1400s John Taylour of Thorndon left 40 pence to the leper hospital of Norwich, 40 pence to that of Eye, 20 pence to that of Bury and 20 pence to that of Ipswich.

In 1463 William Goolding of Eye bequeathed 'to the gild of the Blessed Mary 2s, to the gild of St Peter 2s, to each poor person living in 'le Spytell' 1d.'

The hospital continued until the Dissolution, after which it continued as an almshouse called the Maudlin.

Out in the countryside there were fewer bridge hermits and even fewer bridge chapels but fords and bridges were often marked with stone crosses placed at river or stream bridges. It has been suggested that stone crosses found at river crossings are boundary stones, designed to mark the divisions between the old hundreds. They are usually late Anglo Saxon or early medieval and those bridge sites where such crosses are found act as good indicators of an ancient fording.

In the 17th century the remains of a medieval stone cross could still be seen at the bridge in **Wisset** (called then Wisset Cross). The Suffolk divine and physician Nathaniel Fairfax recorded in 1680 that he remembered seeing it as a boy 'at the north foot of the Cart bridge there, but I think since pulled down'.

In the 1930s members of the Suffolk Institute of Archaeology visited **Great Ashfield** to view an ancient stone cross, standing 10ft 6inches tall, with a carved shaft, possibly dating to as far back as the 11th or 12th century. The Revd H Copinger Hill told members, 'Tradition avers that the Cross once formed a bridge across the stream at the entrance to Great Ashfield Churchyard, but it is more likely that the upper part was used as a stepping-stone across the water.' Discussions ensued as to whether or not the cross had once stood upright and was an Anglo Saxon boundary marker or even a Sanctuary Cross for criminals. The cross had been removed by Lord Thurlow (Lord Chancellor from 1778 to 1792) to Ashfield House in the 1700s. He erected a bridge in its place.

At **Haverhill** a slab of stone has been found in the Stour Brook and is described as a marker stone for a river crossing.

Any markings found on such stones do not always indicate their precise age so it is difficult to ascribe them to a particular custom or period in history.

In the 1530s Henry VIII implemented the Dissolution of the Monasteries. With the demise of the ecclesiastic bridge keepers the business of maintenance and renewal once more entered a state of uncertainty. In 1547, early in the reign of Edward VI, the bridge chapels and their attendant hermits were almost overnight disenfranchised. They were dissolved in the same way as were chantry chapels and converted to other uses, such as lock ups, almshouses or warehouses. Others were dismantled and their fabric re-used. This was often very bad news for those living near to bridges which had hitherto been funded and maintained by a religious community. Such arrangements had usually been preferred to a bridge that belonged to a local worthy who would in due course die and invariably leave an unsympathetic heir or no heir at all.

At **Blythburgh** possession of the priory was granted to Walter Wadelond of Needham Market. In 1537 lead from the roofs was removed and sold; the buildings were robbed of their stone which was taken to repair Blythburgh Bridge. No doubt the bridge was considered unfit for purpose and in a poor state of repair, the prior having been unable to fund the maintenance.

Also in 1537 the Benedictine Priory of St George at **Thetford** was dissolved and granted to Sir Richard Fulmerston (who also benefited by the grant of tolls on the bridge at Brandon) and the Priory buildings converted it into a private home. Sir Richard was closely connected with influential Howard family.

One of Thetford's Nuns Bridges, once the site of the town's ducking stool, has a strange tale connected with it concerning Sir Richard's household and the death of a small child, Lord George Dacre, who succeeded his father as 5th Baron Dacre in 1566 at the age of five. Soon after his father's death, his widowed mother Elizabeth married Thomas Howard, 4th Duke of Norfolk, but died in childbirth soon afterwards. George and his three sisters remained under the care of their step-father, the Duke. However, in 1569 George was sent to Sir Richard Fulmerston's home in Thetford where he died after a fall from a wooden rocking horse.

Perhaps Sir Richard was not popular in Thetford as he was blamed for the death of the six year old George and he had effectively murdered him by removing some pins from the rocking horse thus making it unsafe. The fact that Sir Richard had died two years before the incident, in 1567, appears to have made no difference to his accusers and the ghost of George Dacre riding a headless rocking horse was said to haunt the Nuns Bridge.

Hadleigh's Toppesfield Bridge

Snape Bridge marks an ancient crossing over the River Alde

Plaque and keystone on Eye's Abbey Bridge

Plaque on Brandon Bridge

Modern bridge at the historic Blythburgh crossing

Great Thurlow ford alongside the bridge is still used

Moulton's famous packhorse bridge

The remains of the Cavenham packhorse bridge

Ixworth's Hempyard Bridge under repair in 2016

John Abell's keystone on Nayland Bridge

East Suffolk arms on Southwold's Mights Bridge

The Duke of Suffolk's bridge at Westhorpe

One of the twin bridges in Ipswich over the Stoke Crossing

Sudbury's Ballingdon Bridge marks a very ancient crossing of the River Stour

Bourne Bridge, Ipswich, one of only three Suffolk bridges with pedestrian refuges

One of the Nuns Bridges in Thetford

Mutford's Anglo Saxon crossing is now a busy and complicated junction

Wormingford Bridge on the Suffolk-Essex border

CHAPTER 4

By the middle and late 15th century bridges were an increasingly necessary factor of everyday life and commerce as is witnessed by the number of bequests made to assist with their upkeep. These secular bequests show no distinction between contributions towards the upkeep of bridges and the religious or charitable legacies. Bridge-work in medieval times was one element of communal obligation, a continuation of the aforementioned three 'common burdens' reaching back to the 7th century. There was also the symbolic nature of the bridge in Christian life: after all, bishops and the Pope himself was a *'pontifex'*, a bridge-builder.

In 1462 William Grene of Creeting St Peter left funds toward the 'emendation and reparation' of the bridge at **Earl Stonham** called 'Ladyesbredge'. Mr Grene was obviously a canny testator as the sum of 10 shillings was only to be paid when the work was carried out.

In April of the same year William Chatysle of **Ixworth** left 2 shillings to the reparation of the parish route, which passed over Blackbourne stream and reputed to have been a packhorse bridge. In medieval days the Ixworth watermill, a short distance upstream just beyond the Hempyard Bridge, was run by the Priory of Austin canons when it was the prerogative of religious orders to mill wheat. The priory was established in 1170 and as a result the bridge saw much traffic as a result, right up until the Reformation.

The village was also on the old Roman path known as Peddars Way which further ensured the bridge's importance especially after Ixworth was granted a market in the late 14th century. In 1545 Hewe Baker the Elder left 40 shillings in his will 'to the making of one bridge of lyme and stone at the mylle of Ixworth.' It might, however, be more likely to have been repairs to an existing bridge rather than a new construction.

In September 1464 John Cotelere of Barnham requested that he be buried in the churchyard of the parish of St Martin of Barnham and 'to the emendation of a bridge called Palmers Brygge' he left 6 shillings and 8 pence.

Henry Monnynge of **Norton** near Woolpit died in February 1467/68 and left 40 shillings to the 'emendation of the highway in *Howstrete*' but only if it was needed. To the emendation of the bridge called 'Nortonbrigge', however, he left 'one or two trees'. Presumably it was left to his executors as to how many trees were required and shows the bridge to have been wooden.

In his will of June 1470 John Halle of **Woolpit** ('Wolpet') left funds for 'the reparation of the bridge called *Wolpetbregge* by the way leading to St Edmunds

26 shillings 8 pence' and to the common ways in the town of Woolpit 13 shillings and 4 pence.

Where the local abbey or priory controlled the bridge the standards of repair varied and in the mid 1500s, during the years leading up to the Dissolution of the Monasteries, many were dilapidated and the age-old arguments about who was responsible for the upkeep persisted.

In 1530 the increasing importance given to the state of the nation's bridges was sufficient to bring about the Bridges Act, Section One of which empowered Justices of the Peace to enquire into matters of broken bridges and to order their repair or rebuilding. This was usually at the expense of those who were already deemed to be responsible for their maintenance which is predictably where many disagreements arose. Section Two took account of this and decreed that where those responsible could not be determined, the expense would fall on the inhabitants of the town where the bridge was situated. If the bridge lay outside a town then the burden would fall on the shire as a whole. Thus, for the first time, precise responsibility for bridge maintenance was enshrined in law. Whether or not it was popular with those on whom the burden fell is another matter.

In 1555 a further statute was enacted which required that during Easter Week each parish should elect two surveyors of the highways and bridges. These surveyors should name in church four particular days before the Feast of St John the Baptist when men and materials would be sent out to repair and maintain the King's highway. Since it was virtually impossible for such an order to be policed its effects were spasmodic.

There was, though, still room for those who considered bridges a private matter and are best described as 'mercantile gentry'. During the 16th century it is logical that town benefactors included merchants and clothiers whose business concerns relied heavily on good access to their markets. One such was John Abell a wealthy clothier of **Nayland**, who died in 1523. In his will he provided for the 'perpetual maintenance' of the Anchor Bridge, the keystone to carry a hand bell with an 'A' atop, a pun on his name. The money was to come from the proceeds of his farm and land at Layer Breton near Colchester in Essex. Unusually, his wishes were carried out not perhaps in perpetuity, but until 1704 when responsibility passed to a trust.

Abell's wooden bridge was replaced by a brick bridge in 1775 when the hand bell keystones were incorporated on the up and downstream parapets of the new bridge and the date 1775 added. These two keystones were again incorporated in the structure when the present bridge was built in the 1950s.

When the stone bridge replaced the old wooden one it was observed that although it had served the town well for some years, it was in a dilapidated state. Part of the problem was that the bridge straddled the county boundary, formed by the river Stour; accountability was shared between Suffolk and Essex and thus invariably disputed even when paid for by the Abell family.

The bridge is called variously Abell's Bridge, Plod Bridge, Anchor Bridge (because it stands near the Anchor Inn) and Bell Bridge.

Another somewhat less successful town benefactor was Roger Barrow of **Clare** who in 1598 engaged in Chancery Proceedings with several Clare inhabitants over his fiscal responsibility for the bridge.

The geography of Clare meant that it has had several bridges over the centuries. The medieval approach to the town is described by the 1920s Clare historian, Gladys Thornton:

> Coming from Yeldham travellers crossed the Stour river by a small wooden bridge, and turned right into the Nethergate, passing a few dwellings such as the Stonehall. At the top of the street Mill Lane led to a great stone arched bridge which linked the castle and the Austin friary, and here a water mill was used for malt. At the north end of the market the road turned right into Cavendish, skirting the castle bailey and crossing the Chilton stream by a bridge known as the Pysenebregge.

Pysenebregge was later renamed Baybridge and some years afterwards became a turnpike crossing.

Roger Barrow was seemingly a leading light in 16th century Clare. He was a grocer, yeoman and possibly an innkeeper (as he bequeathed the Green Dragon to his wife) and generally took part in town life including taking on the duties of church warden. However in 1597 Barrow decided that he had spent too much money on the town and instituted a suit in Chancery against thirteen of the town's residents in an attempt to recover his costs. Among his many expenses, he claimed, was a large quantity of timber destined for the repair of the town bridges. His opponents, however, claimed that among other misdemeanours he had sold some of the timber and only repaired part of the Baybridge.

The town bridges were said to be 'altogether overturned and carried away with water' so that Her Majesty's subjects were greatly hindered in their passage thereby. The complainants also declared that such was the state of Baybridge that 'the water could have no passage … so that the next flood that came the said bridge was carried away.'

It is probable that Roger Barrow had overstretched himself financially and was juggling his various interests, using Peter to pay Paul. It is very likely he did sell the bridge timber but it could not be proved and the case was dropped.

In the decades following the 1530 Bridges Act it was easier to apportion responsibility for bridges and allow local authorities to intervene. In 1570, for example, in spite of monetary help from the Aldeburgh Corporation and the Orford Corporation, the inhabitants of **Snape** were fined for the disrepair of their bridge. Some years later Snape residents made an agreement that they should repair the northern end and those in Tunstall should maintain the southern end (Snape Maltings is actually in Tunstall parish). This proved predictably problematic and many years later the inhabitants of both villages were again fined for the bridge's decay.

Snape Bridge was of strategic importance as being on the main route to London and over time would have been subjected to sustained pressure from increased traffic. It is the point where the River Alde becomes tidal and, while it is not possible to know when the first bridge was thrown across its waters, there is no doubt that a crossing of some kind has been there since time immemorial. It is featured on all the 16th and 17th century maps and Robert Reyce mentioned it in his *Breviary*.

At **Moulton** the cost of repairing the bridges was a heavy burden on such a small town and as a result a charity known as 'The Church and Bridge Estate' was set up in 1532. Rents from estates at King's Fen at Freckenham and miscellaneous estates in and around Moulton provided income for 'reparation of the church and of the bridges in the parish.'

In 1598 at **Eye** the tenant of Eye Priory had to maintain 'the causeway leading towards the church and five timber bridges namely the three Abbey Bridges, Largate Bridge and Botsford Bridge.' This was one example among many where an existing person could be found and forced to carry out repairs to the bridge fabric, the tenant having the dubious honour of inheriting the priory's pre-Reformation responsibility.

In 1618 the Town Book of **Brockdish** (Norfolk) recorded that 17 shillings were laid out for the repair of the Brockdish part of the bridge leading over the river to **Syleham** in Suffolk. The bridge had recently collapsed into the river 'through the negligence of both the parishes, though it was of equal service to both.' In the end the two villages went half-and-half and the bridge was duly reinstated.

The bridges at **Southwold** and **Reydon** were also crucial to both places and during the 1470s bequests were still made to 'the great bridge' at Southwold, the bridge across Buss Creek where Mights Bridge now stands. The first Mights Bridge seems to have been built in 1256 when the town was allowed to bridge the creek to facilitate traffic to and from their market. There were undoubtedly generations of bridges at that point and they have all carried the same original name, Mights.

The bridge known today as Wolsey Bridge Sluice (Woolsey Bridge on Hodskinson's 1783 map) was once called Wevylsee Bridge and crosses the River Wang on the Halesworth road between **Reydon** and **Bulcamp**. Tradition has it that Ipswich-born Cardinal Thomas Wolsey was a benefactor to the parish of Reydon and that he paid for considerable improvements to Wevylsee Bridge and associated causeways, thus inspiring its change of name.

Although Wolsey was said to be benevolent towards his home county it did not pay to fall foul of him. A seven-foot high drawbridge was employed at the early 16th century semi-moated hall in **Little Saxham** (now demolished). The hall was approached by a causeway with a gatehouse and the moats were dug both deep and wide. The enhanced security might be due to the fact that the builder, Thomas Lucas, was at loggerheads with Cardinal Wolsey. If so his apprehension was justified as Wolsey later imprisoned him in the Tower 'for speaking scandalous words of the Lord Cardinal'.

By the 17th century bridges appearing in wills changed to become means of identifying the whereabouts of property rather than as recipients of bequests and there were few, if any, public benefactors prepared to take on repair and renewal costs. In 1632 John Gainsford of **Icklingham** left 'land abutting on the highway to Lackford bridge' as well as '3 rods of infield land abutting on the highway to Lackford bridge'.

Also in 1632 Robert Howe of **Sudbury** left his 'loving and well beloved wife' Judith his messuage and lands in Sudbury 'near the bridge foot sometime called Ballingdon Bridge'.

In **Ipswich** the condition of the bridge at Stoke was too important to be left to chance. It was always the object of much consideration within the town. Prominent townsmen found it expedient to endow the crossing when necessary throughout Tudor times. The 19th century writer and historian John Wodderspoon pointed out many entries in respect to it in the town books such as:

> In the 13th Henry VI (1435) John de Caldwell offered to make a bridge at Stoke at his own cost, provided the inhabitants would pay pontage.

Carriers had long been required to pay a toll over Stoke Bridge but in 1478 an order was made that carts should not go over the bridge and that it should be kept locked. Increased vehicular traffic meant that it had become unsafe.

During the reign of Henry VII a new bridge was built and an order was made that all carters using Stoke Bridge ('lately built') 'shall pay towards the repairing and maintaining of the same, every Burgess a half-penny, and every foreigner a penny, provided that none crossed who could pass the ford.'

By using the ford, wear and tear could be saved on the fabric of the main bridge.

In the first year of Elizabeth I's reign, the corporation made provision for all future repairs of the bridge, large quantities of wood to be sourced from Whitton and Holbrook for the 'building of Stoke Bridge'. However, by 1610 a stone bridge, rather than a wooden one, is seen on John Speed's town map where it is shown to have a house built on one of the piers, probably a toll house from which the bridge master would conduct his duties. It is likely that the stone bridge was in place by 1594 when an order was given that both the Stoke Bridge and Handford Bridge should be 'made sufficient for men, horses and carts'. What is certain in all this is that there was still a ford close to Stoke Bridge that could be used and was, indeed, still used until early in the 19th century.

Speed's map also shows that shipping could not go beyond the bridge and it would have been necessary to transfer cargoes onto smaller boats to go further up river. What is clear, however, from the account of the town corporation is that repairs were constantly needed at the Stoke, Handford and Friars bridges during the 16th century. The side posts of Stoke Bridge were frequently in need of repair.

Bridges that left no room for financial disputes are the many hundreds thrown across the county's moats. It is estimated that there are still between five and six hundred moated properties in Suffolk and each one would have one if not two bridges. Many date back to the 16th century although that at **Wingfield** was built in the 14th century.

The Hall at **Playford** has bridges over its moat and during Tudor times was one of the most eminent mansion houses in England. Henry VIII was once entertained at Playford Hall. The bridges and their associated walls have been dated as late 16th century and the moat itself runs along the south bank of the River Fynn.

Moats have been popular since the 13th and 14th century but opinions about their *raison d'être* are various. It was usually thought to be for defence purposes, against invading forces, or purely decorative or symbolic. A study of moats carried out in the 1970s by the historian Colin Platt, concluded that they were intended as defences against marauding peasants in periods of agrarian unrest. They also successfully prevented livestock from straying and in the Suffolk claylands moats provided drainage and retained winter water for the summer months.

One of the most attractive of the county's moat bridges is the Tudor one at **Westhorpe** which has close associations with Henry VIII's younger sister, Mary Tudor, wife of Louis XII of France. Very soon after her French husband died she married Charles Brandon, 1st Duke of Suffolk, as his third wife thus incurring the wrath of her brother, Henry. She had married without permission of the sovereign and was, therefore, guilty of treason. Charles, his closest friend, was thus accused of disloyalty.

Eventually Brandon bought his brother in law's favour with a large amount of money and jewels. In truth, Henry had been more interested in the gold plate, jewels and considerable dowry that Louis XII had conferred on Mary on their marriage and made it a condition of his agreement to the marriage that it should all be handed over to him. After considerable negotiating conducted by Cardinal Wolsey on behalf of the king, a deal was struck and the couple allowed to return to England from France.

Charles Brandon built Westhorpe Hall and, as the Duke and Duchess of Suffolk, set up home there: they raised their three children at Westhorpe and Mary died there in 1533. The mansion was pulled down in the 1760s and all that is left is the three-arched brick bridge built across the moat. Although it is classed as Tudor, since most of it was built in Brandon's time there, it is believed to have been built much earlier but allowed to fall in disrepair. The ruined remnants of a footbridge were once discovered, which would have formed a crossing over the eastern arm of the moat, but these have now disappeared.

The three-arched bridge is faced with brick and originally had parapets with stone pillars to support figures of heraldic beasts. On the south side it has the remains of a frieze of terracotta panels with Brandon's arms and the head of a lion. The Duke and Duchess of Suffolk would have crossed this bridge numerous times during their residence at Westhorpe and no doubt Mary's body passed over it on her way to burial in the Monastery at Bury St Edmunds.

Another Suffolk bridge that played a small but important role in the course of British Tudor history is another moat bridge, this one at Framlingham Castle. During the dramatic events of 1553, when Edward VI died and the succession became uncertain, Mary Tudor was staying in Kenninghall (Norfolk). On hearing the news she and her entourage immediately made their way to **Framlingham** where she stayed long enough to amass support for her claim to the throne. Only the piers remain now of the Tudor bridge that once led directly to the Inner Court of the castle and to the lodging which by local tradition was occupied by Princess (soon to be Queen) Mary. At Framlingham she attracted many more followers who saw her as the true queen, rather than the unfortunate Lady Jane Grey (granddaughter of Mary Tudor of Westhorpe) whom her brother Edward VI had named as his heir. Over that bridge Mary rode out into the Suffolk countryside with her loyal followers to claim her throne, arriving in London on 7 August where she triumphed. The bridge piers stand close to the seventh tower.

A Tudor bridge at **Sudbury** was built in 1521 but was constantly under repair. It spanned the River Stour where the Ballingdon bridge is today but collapsed in 1594 under the pressure of a great flood. The 1521 bridge was about 3m wide with eight arches, each one about 4 m across. Archaeological excavations carried out on the site discovered extensive brick and mortar remains from the 1521 bridge, as well as carved limestones from the 12th century bridge. Between 1586 and 1805 there were at least four different timber bridges on the same site. From 1549 until 1888, Justices of the Peace from Suffolk and Essex ordered repairs to be undertaken from the part of the bridge under the jurisdiction of each county.

A year after Mary came to the throne the Mutford Bridge at **Oulton Broad** was built and financed by Katherine Mayde although the bridge was subsequently swept away several times by winter storms. In 1660 a breakwater was constructed at **Lowestoft** and a new wooden bridge built at the Mutford crossing which remained in place until 1760. It was the only way into Lowestoft for wheeled traffic from the south and as a consequence took on heightened importance. The Mutford crossing is one of those which appear on Christopher Saxton's 1576 county maps.

Christopher Saxton, born around 1540, was an early cartographer who produced an *Atlas of the Counties of England and Wales*, published in 1579. It was the first such atlas and was carried out under the patronage of the Queen and Thomas Seckford, a lawyer and Member of Parliament for Ipswich. Seckford was born in Suffolk in 1515 and rose up the ranks to become an official in Elizabeth I's court. In 1564 she sold him the Manor of Woodbridge and he became a benefactor to the town. He financed Saxton's survey which began in 1574. Five years later the *Atlas* was complete and contained 35 maps each bearing the arms of the Queen and Thomas Seckford.

Saxton employed the latest advances in engraved copper plates and improved surveying technology. However, in the burgeoning profession of cartography it was the road systems that took priority and bridges were very much taken for granted.

Throughout history the East Anglian coast has been vulnerable to invasion from the east and in 1588 it came in the shape of the Spanish Armada. Philip II of Spain sent an invasion fleet comprising around 138 vessels with 7,000 seamen, 17,000 soldiers and siege equipment with the intention of landing on the east coast. Trouble between Spain and England had been brewing for some time in the matter of trade and religion and Elizabeth prepared to strengthen coastal defences.

A map of 1588 (a copy of which is in Southwold Museum) shows that a three-pronged fort was planned for Southwold to be cited at Easton Bavents down river of Mights Bridge with a clear view of the open sea. However, although a bridge fort had certainly been proposed and mapped, there is no evidence that it was ever constructed.

The idea of a fort at Southwold was not a new one. In 1262 Richard de Clare applied to Henry III for permission to build a fort, or castle, but although it was granted there is nothing to say where it was built or even if it was built at all.

In spite of legislation and the ever-increasing commercial logic for good transport links across the county the overall state of the bridges was poor. In 1618 Robert Reyce of Preston St Mary published *Suffolk in the XVIIth Century* although it had been written fifteen years earlier, in 1603. He provides a contemporary view on bridges and wrote:

> Neither can I add any better report of our bridges, which suffering their decay in time throw the continuall violence of the floods, would receive a totall distruction (whilst every one denies the charge putting it off from himselfe) to the great interruption of the common entercourse, if the different Magistrate by the life and execution of good lawes did not enforce a repaire of the same.

Robert Reyce was just one of a breed of new and inquisitive writers and topographers who would begin to record what might be thought of as the minutiae of Suffolk life but what is now regarded as a hugely important contribution to history, bridge history in particular.

No one wanted to pay for the bridges and the law was not always enforced. The old obligation of '*pontium reparatio*', where the repair of bridges was defined and declared by Henry VIII's Statue of Bridges, could still be sidestepped without too much trouble. As a consequence many bridges that still relied on benefactors often suffered when said benefactors defaulted on maintenance, usually due to a lack of money brought about by a downturn in fortunes.

For example, the bridge between **Stoke Ash** and **Wetheringsett** was often in need of repair but in 1599 it was the local people who paid. The inhabitants of Stoke Ash paid three quarters of the costs and Wetheringsett the remaining quarter but it is doubtful that it had been their responsibility in former days.

In 1600 it was the town of **Newmarket** which paid for the new bridge over the River Kennett. Local legend has it that it was for the benefit of James I who passed through Newmarket on his way to visit Thetford where he took part in falconry activities on which he was very keen. The river crossing was thought to be inadequate and the town forced to bear the cost of a new bridge.

There had been a crossing of the Kennett at that point since 1161 which is thought to have contributed to establishing the 'new market' in around 1200. Whether or not the bridge was built for the king it had the added advantage of opening up the route through Newmarket to Thetford and beyond and eventually became so busy that it was turnpiked.

By the early 1600s the ease and convenience of good road bridges began to be the norm. Until, that is, a bridge was swept away and travellers were reminded why ancient routes across the old fords were still vital to the county as a whole.

In 1610, the year that Christopher Saxton died a new cartographer and historian came on the scene. John Speed, born in 1552, worked in London and came to the attention of Elizabeth I. Speed's map of Suffolk, published in 1610 and subsequent editions, shows the county with its hundreds and includes a town plan of Ipswich.

The Ipswich map clearly defines the Stoke crossing and was described by the 19th century historian John Wodderspoon as being constructed of stone 'with a house built upon one of the piers, either for the convenience of levying pontage, or as a defence.'

The bridge on Speed's map, thought Wodderspoon, seemed to have been erected in a 'ponderous' style, and the arches showed a low and pointed character.

Friars (Friers) Bridge is also shown close to the Franciscan Friary on the north bank of the River Orwell that bridged a stream that joined the River Gipping.

In Elizabethan times the three main bridges of Ipswich were in constant need of repair and their maintenance was one of the chief drains on Corporation finances. New planks and posts were frequently needed at Friars (Friers) Bridge and at 'Handforthe' Bridge. In 1559 the latter needed new planks although in 1571 the reference is to 'amendinge the little bridge' at the Handford crossing which raises the possibility of more than one bridge at the crossing. In 1586 money was paid to Jhon Braye for 'a pece of tymber to Handford Brege to mend it withall' and in the same year yet more outlay to Jhon Whaytes 'for 9 foet tymber for the breg 9d, for plank to mend the breg 8d, for naylles 2d and for workmanship 3d.'

In 1577 huge repairs were carried out at Stoke 'repayring of the decaye at Stoke Bredge'. New side posts were required and large stones were lifted from the river to place 'them to kepe the cartes from the railes.' In 1593 extensive repairs were carried out at 'Handfourd' and 'Stooke' bridges which required planks of wood, gravel, several pieces of timber to make joists, timber for posts and labour costs. More charges were incurred in felling the wood and carting it from Whitton.

Quite what state the Melford Bridge crossing at **Thetford** was in during the 1600s is unknown but probably in poor repair since Sir John Wodehouse, 4th Baronet and Member of Parliament, paid for the two-arch red brick and stone bridge to be rebuilt over the River Thet in 1697. Sir John was at one time the Recorder for Thetford and on the upstream side of the bridge the pilaster has a stone plaque bearing the Wodehouse arms and a Latin inscription giving Sir John's name and date of construction.

Part of the Melford Bridges lies in the parish of **Brettenham** and at some point the bridge was extended and another arch added. Although it is now spelled Melford, it is thought that the earlier name for the crossing was Mill Ford Bridge which suggests that a medieval watermill stood close by the crossing.

Suffolk's bridges did not, on the whole, enter the 18th century in a particularly good state but from then onwards travellers and cartographers began to chart their history not individually, but as part of recognised routes in an increasingly urbanised landscape. More traffic, particularly the use of vehicular transport, increased the burden on parishes and townships to such an extent that a solution had to be found: it came in the form of turnpike trusts which were empowered to levy tolls for the upkeep of named roads and their attendant bridges. Towards the end of the 16th century and from the 17th century onwards the ever greater weight and size of carts and carriages inflicted ever more damage on the wooden structures and hurried the introduction of stone replacements. As the volume of traffic increased the maintenance of bridges became paramount and who paid for them became ever more crucial to the prosperity of Suffolk.

SELECT BIBLIOGRAPHY

Bailey, Mark *Mediaeval Suffolk* (2007)

Cook, Olive *Suffolk* (1948)

Cooper, Alan *Bridges, Law and Power in Medieval England, 700-1400* (2006)

Cruickshank, Dan *Bridges: Heroic Designs that Changed the World* (2010)

Harper-Bill, Christopher *The Dodnash Priory Charters* (1998)

Hodskinson, Joseph *Hodskinson's Map of Suffolk in 1783* (2003 edition)

Jervoise, E *The Ancient Bridges of Mid and Eastern England* (1932)

Kirby, John *The Suffolk Traveller, 1735* (Suffolk Records Society 2004)

Middleton-Stewart, Judith *Inward Purity and Outward Splendour* (2001)

Reyce, Robert *Suffolk in the XVII Century* (1618)

Scarfe, Norman *The Suffolk Landscape* (2002)

Suckling, Alfred *The History and Antiquities of the County of Suffolk Volumes 1 and 2* (1846)

Wallace, Doreen *East Anglia* (1939)

Wodderspoon, John *Memorials of the Ancient Town of Ipswich* (1850)

How to find Suffolk Towns and Villages Suffolk Family History Society (2000)

The Suffolk Village Book Suffolk Federation of Women's Institutes (1991)

INDEX

Alpheton	5	Combs Ford	4
Barnby	8	Cookley	5
Beccles	19, 33-34	Debenham	5, 7
Bentley	33	Earl Stonham	49
Billingford	4	East Bergholt	33
Blythburgh	15, 38, 42	Eastbridge	12, 36
Boxford	5, 27	Eye	11, 19, 37, 41, 52
Bridge Street	11	Fordley	4
Brandon	6, 15, 20, 34, 41	Framlingham	10, 56
Brockford	4, 7	Glevering	13
Bruisyard	13	Glemsford	4, 7
Bungay	35	Gosford	9
Bures	19	Grundisburgh	5
Bury St Edmunds	16-17, 31, 37	Hadleigh	17, 36, 40
Catttawade	4, 6, 32	Haverhill	19, 38
Cavenham	23, 35, 43	Hawkeswade	3
Charsfield	1	Horsewade	4
Chelsworth	26	Hoxne	1, 8, 30
Chillesford	4	Icklingham	35, 53
Clare	6, 16, 51	Ipswich	4, 5, 7, 10, 12, 17-19, 46, 47, 53-54, 58
Coddenham	7		
		Ixworth	1, 24, 44, 49

Kentford	4, 6, 23	Stoke Ash	6, 13, 57
Kettleburgh	13, 28	Stoke by Clare	19
Kersey	5	Stratford St Andrew	6
Knettishall	2, 9	Stratford St Mary	6, 36
Lackford	4	Sudbury	14-15, 36, 46, 53, 56
Little Cornard	19	Syleham	1, 52
Lowestoft	56	Thelnetham	4
Moulton	21-23, 43, 52	Thetford	1, 8, 24, 38, 47, 59
Mutford	9, 48	Ufford	3, 26
Norton	49	Walsham le Willows	1
Nayland	44, 50	Walsingham Shrine	15
Newmarket	57	West Stow	4
North Cove	4	Westhorpe	45, 55
Olaves, St	6, 20-21, 31	Wetheringsett	59
Oulton Broad	9, 56	Wilford	9, 30
Playford	4, 54	Wissett	9, 37
Rendham	2	Withersfield	7
Risbridge Hundred	12	Witnesham	12
Risby	23	Woodbridge	11
Semer	33	Wormingford	48
Sibton	16, 29	Woolpit	49
Snape	15-16, 40, 52	Yoxford	4
Southwold	11, 45, 52-53		